HARCOURT

Math

Practice
Workbook

Grade K

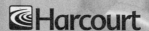 Harcourt

Orlando   Austin   Chicago   New York   Toronto   London   San Diego

Visit *The Learning Site!*
www.harcourtschool.com

CONTENTS

Top, Middle, Bottom

© Harcourt

Color the bear on the top shelf. Color the doll on the middle shelf.
Color the ball on the bottom shelf.

Name _____

In, Out

- Circle the dog that is in the doghouse.
- Circle the chick that is out of the egg.
- ★ Circle the girl who is in the house.
- ♥ Circle the duck that is out of the pond.

PW2 Practice

Above, Below, Over, Under

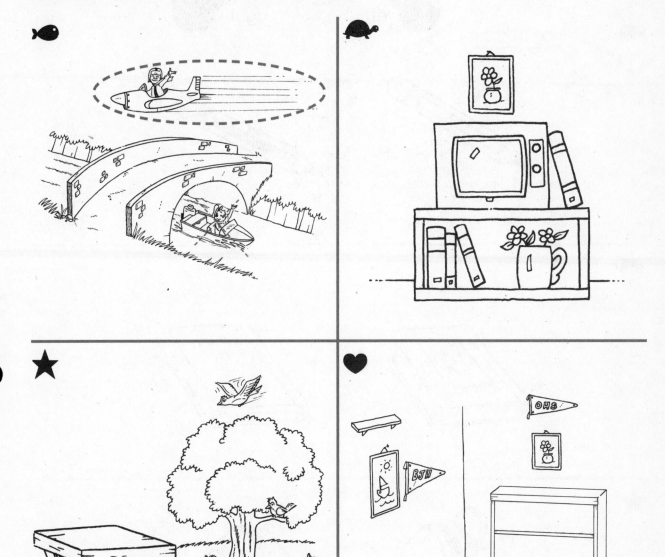

🐟 Circle the man over the bridge. Mark an X on the man under the bridge.

🐢 Circle the flowers above the television. Mark an X on a book below the television.

★ Circle the bird over the tree. Mark an X on the cat under the table.

♥ Circle the pennant above the picture. Mark an X on the shelf below the picture.

Name _____

Left, Right

 left   right

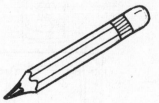   

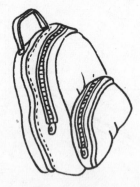   

🐟 Circle the pencil on the left.
🐢 Circle the book on the right.
★ Circle the backpack on the left.

PW4 Practice

Problem Solving Skill: Use a Picture

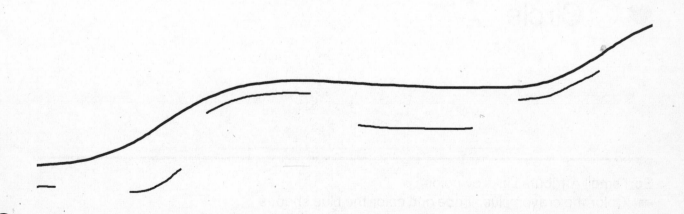

Use red to color the middle tube. Use yellow to color the shovel below the table.
Use green to color the sandcastle on the left.

Name _____

Sort by Color or Shape

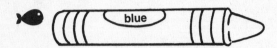

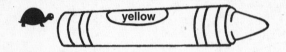

★ ☐ Square

♥ ○ Circle

Sort small Attribute Links by color.
🐟 Color the crayon blue. Trace and color the blue shapes.
🐢 Color the crayon yellow. Trace and color the yellow shapes.
Sort small Attribute Links by shape.
★ Trace and color the square shapes.
♥ Trace and color the circle shapes.

Name _____

Sort by Size or Kind

Circle the objects that are the same size.
Circle the objects that are the same kind.

Make a Concrete Graph

Red and Yellow Counters

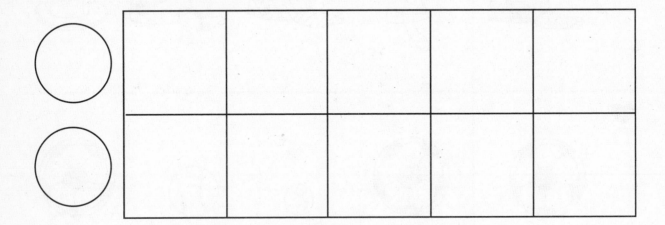

- 🐟 Place a handful of counters in the work area. Sort by color.
- 🐢 Color the counter in the top row of the graph red and the counter in the bottom row of the graph yellow. Move the counters to the graph. What does this graph show?

© Harcourt

Name _____

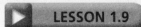

Problem Solving Strategy • Use Logical Reasoning

🐟

🐢

★

♥

 🐢 ★ ♥ Mark an **X** on the object that does not belong.

Name _____

Movement Patterns

★

🐟 🐢 ★ Act out the pattern. Say the pattern as you act out each part. Circle what you would most likely do next.

Name _____

Read and Copy Simple Patterns

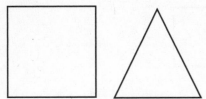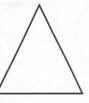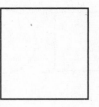

- -
_____

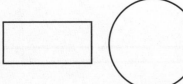

- -
_____

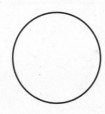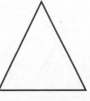

- -
_____

Read the pattern. Draw to copy the pattern.

Name _____

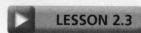

Copy and Extend Patterns

 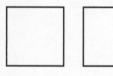

 Draw the shapes to copy the pattern and show what most likely comes next.

PW12 Practice

© Harcourt

Name _____

Predict and Extend Patterns

What pictures do you think come next? Draw the pictures to show what most likely comes next.

© Harcourt

Name _____

Problem Solving Skill: Transfer a Pattern

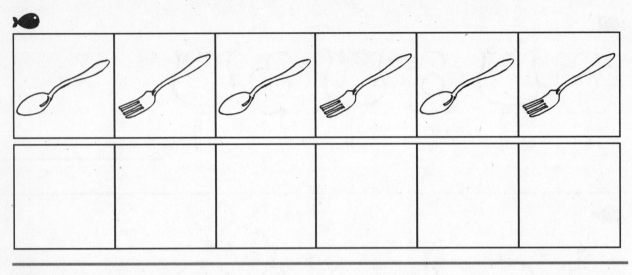

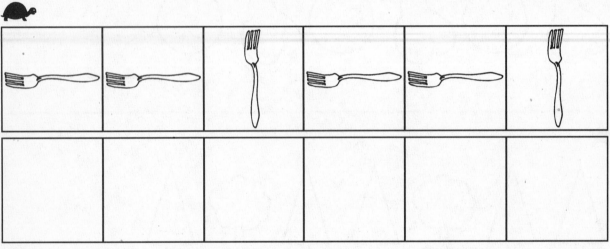

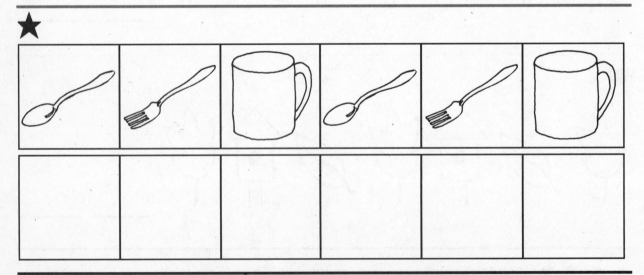

Use counters or act out to show the same pattern. Draw the pattern.

Understand a Pattern

♥

🐟 🐢 ★ ♥   Read the pattern. Circle the part that repeats again and again.

Name _____

Create a Pattern

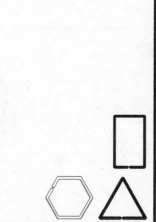

 Use connecting cubes to make your own pattern. Draw your pattern.

 Use Attribute Links to make your own pattern. Draw your pattern.

Name _____

Problem Solving Skill: Use a Pattern

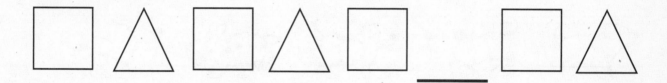

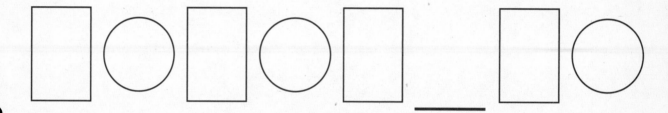

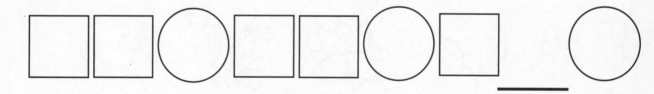

Read the pattern. Tell what part repeats again and again. Draw the missing shape in the pattern.

© Harcourt

Name _____

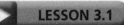

Equal Groups

 Draw a bone for each dog.

 Draw a pot for each plant.

Name _____

More

Draw lines to match the animals in the two groups. Compare the groups. Circle the group that has more.

Fewer

★

  ★ Draw lines to match the animals in the two groups. Compare the groups.
Circle the group that has fewer.

Name _____

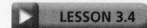

Problem Solving Strategy: Make A Graph

Which Group Has More?

Yellow	Red

Put a handful of counters in the bowl. Are there more red counters or more yellow counters?
Move the counters to the graph. Draw and color. Circle the column with more counters.

© Harcourt

Name _____

One, Two, Three, Four

1 🍎 apple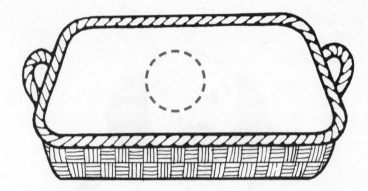

2 🍊 oranges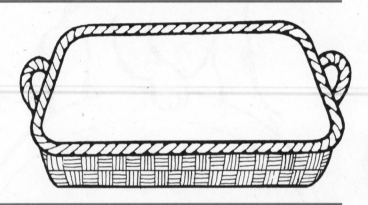

3 🍌 bananas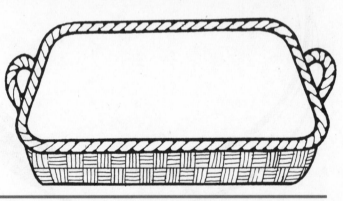

4 🍓 strawberries

🐟 🐢 ★ ♥ Read the number. Draw that many pieces of fruit in the basket.

Five

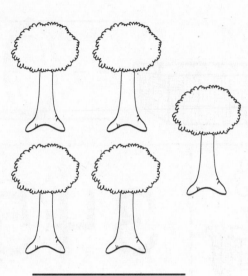

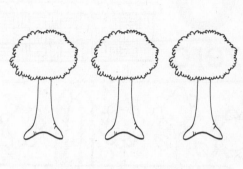

★

♥

Zero

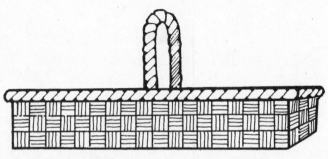

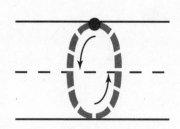

0

zero

3 4 ⑤

★

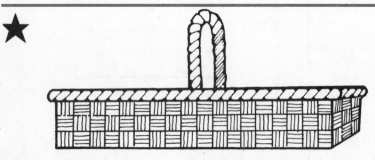

0 1 2

♥

1 2 3

✿

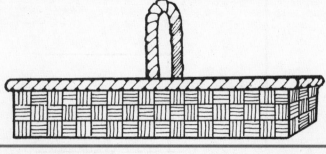

0 1 2

 Read the number and look at the picture. Trace the number.

🐢 ★ ♥ ✿ Circle the number that tells how many flowers are in each basket.

Before and After on a Number Line

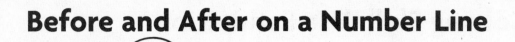

0 1 ---- 3 4 5

0 ---- 2 3 4 ----

---- 1 2 ---- 4 5

Write the number that is before 3.

Write the number that is before 2. Write the number that is after 4.

★ Write the number that is before 1. Write the number that is after 2.

Name _____

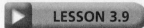

Problem Solving Skill: Use Estimation

Look at the fish at the top of the page. Without counting, use red to circle the fish that have more than five spots. Use blue to circle the fish that have fewer than five spots.

Six and Seven

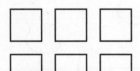

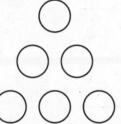

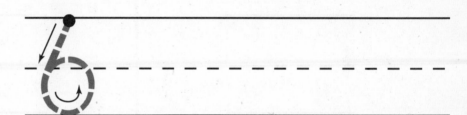

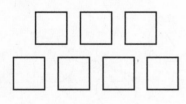

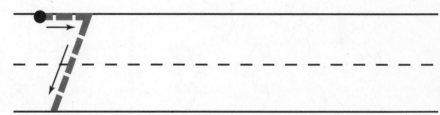

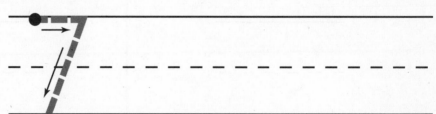

  ★ ♥ Count the shapes in the group. Write the number.

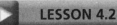

Eight and Nine

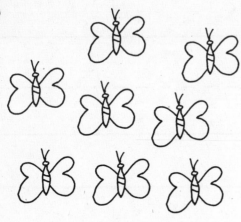

_____

- - - - - - -

_____

_____

- - - - - - -

_____

★

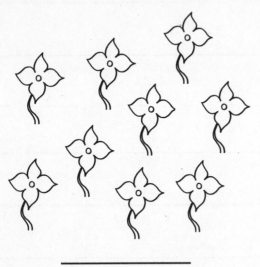

_____

- - - - - - -

_____

♥

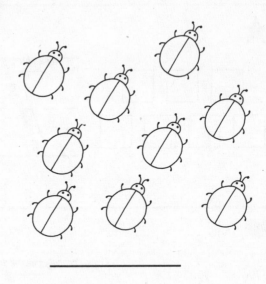

_____

- - - - - - -

_____

 ★ ♥ Count. Write the number.

© Harcourt

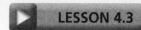

Ten

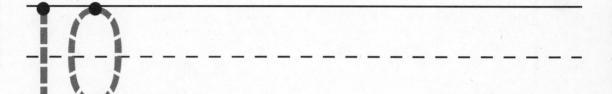

8   9   10

★

8   9   10

♥

8   9   10

🐟 Write the number 10.

🐢 ★ ♥   Count the bees. Circle the number that tells how many. Write the number.

© Harcourt

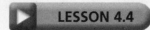

Problem Solving Strategy • Make a Model

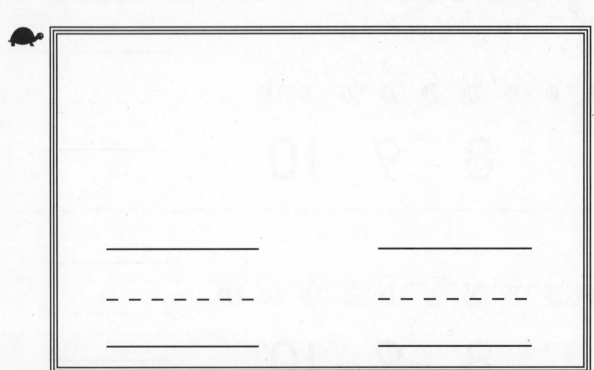

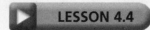 Connect ten cubes. Then make one break to get two groups—one with more cubes than the other. Put the groups on the work space. For each group, draw the cubes and write the number. Circle the number that is greater. Use different numbers to make ten each time.

Name _____

Before and After on a Number Line

🐟

←————|————|————|————|————|————→

6 ____ 8 9 10

🐢

←————|————|————|————|————|————→

____ 7 8 9 ____

★

←————|————|————|————|————|————→

6 7 ____ 10

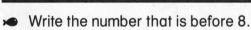

🐟 Write the number that is before 8.
🐢 Write the number that is before 7. Write the number that is after 9.
★ Write the number that is after 7. Write the number that is before 10.

Name _____

Write Numbers 0 to 10

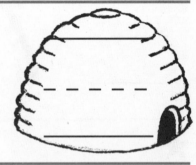

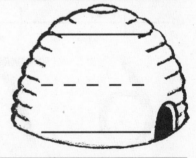

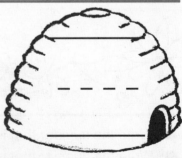

  ★ ♥ ✿ .Count the bees and write the number.

Name _____

Missing Numbers

 5   6   ▢   8   9 | 4 |
|---|
| 7 |

🐢 6   7   8   9   ▢ | 6 |
|---|
| 10 |

★ ▢   7   6   5   4 | 8 |
|---|
| 4 |

♥ 10   ▢   8   7   6 | 9 |
|---|
| 5 |

✿ ▢   7   8   9   10 | 8 |
|---|
| 6 |

© Harcourt

🐟 🐢 ★ ♥ ✿ Circle the missing number.

Name _____

Problem Solving Skill • Use Data from a Graph

Our Pets

_____

Place a connecting cube on each cat in the picture. Move the cubes to the cat row on the graph. Now do the same thing for the dogs. Count the cubes in each row and write the numbers.
Are there more cats or more dogs? Circle the number that is greater.

Sort Solid Figures

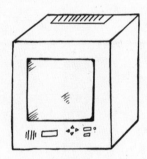  

Use blue to circle the objects shaped like spheres.
Use red to circle the objects shaped like cones.
Use yellow to circle the objects shaped like cubes.
Use green to circle the objects shaped like cylinders.

Name _____

Move Solid Figures

 Circle the shapes that slide.
 Circle the shape that stacks.
★ Circle the shapes that roll.
♥ Circle the shapes that roll and slide.

PW36   Practice

Name _____

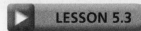

Problem Solving Skill • Use Visual Thinking

🐟

🐢

★

♥

🐟 🐢 ★ ♥ Look at the object at the beginning of the row. Circle the outline that matches the shape of the object.

© Harcourt

Sort Plane Shapes

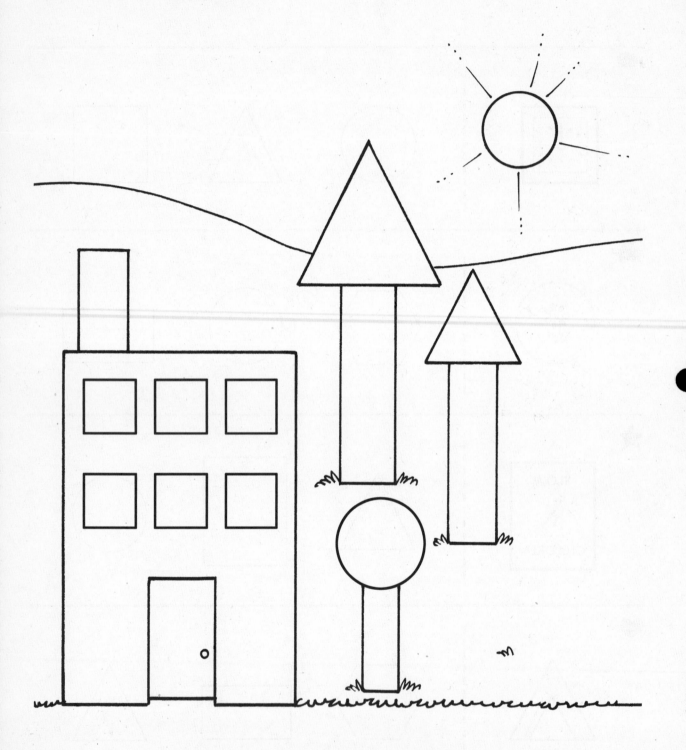

Color the circles yellow. Color the squares blue. Color the rectangles brown.
Color the triangles green.

Name _____

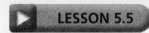

Plane Shapes in Different Positions

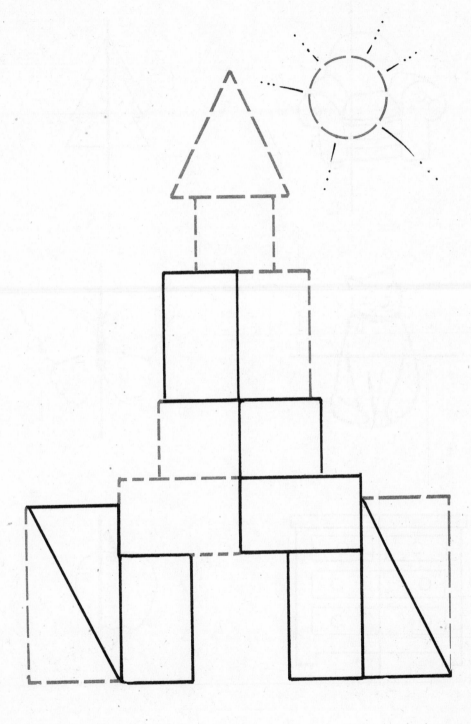

© Harcourt

Use plane shapes to finish the puzzle. Use the same color to color the shapes that are alike.

Symmetry

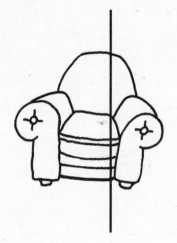

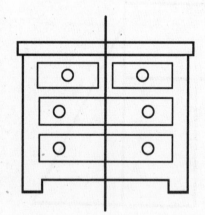

Circle the objects that have a line that divides them into two matching parts.

PW40   **Practice**

Equal Parts

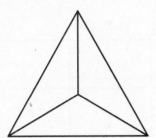 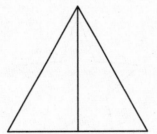 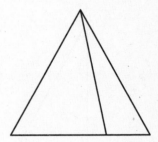

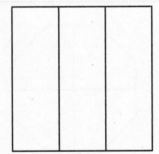 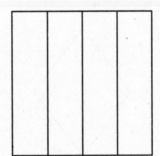 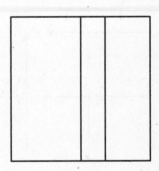

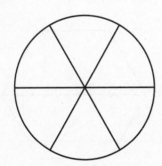 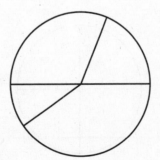 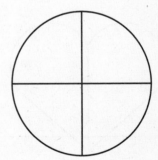

 Circle the shape that is divided into two equal parts.
 Circle the shape that is divided into three equal parts.
★ Circle the shape that is divided into four equal parts.

Problem Solving Strategy • Make a Model

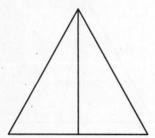  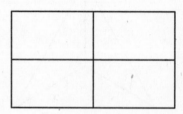

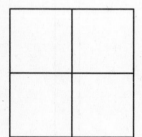 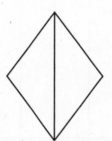 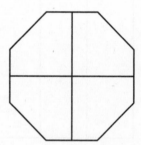

★

  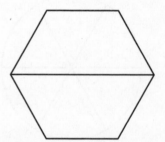

🐟 🐢 ★ Use green to color one part of each shape. Use red to circle the shapes that show one half. Use blue to circle the shapes that show one fourth.

PW42 Practice

Name _____

Problem Solving Strategy: Make a Model

 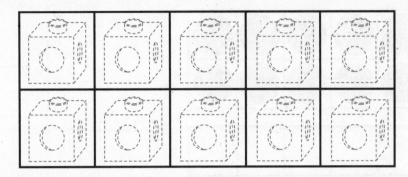 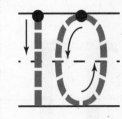

 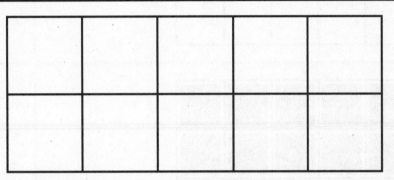

★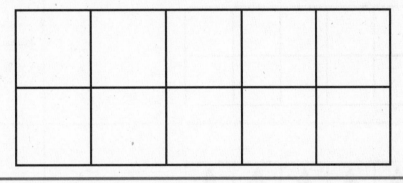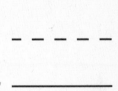

 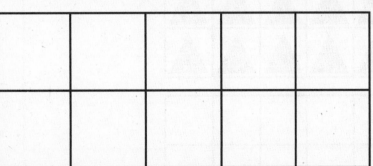

 Place ten cubes on the ten frame. Trace the number.

🐢 Use cubes to model the number that is one more than nine. Write the number.

★ Use cubes to model the number that is one less than ten. Write the number.

♥ Use cubes to model the number that is two more than eight. Write the number.

11, 12, 13

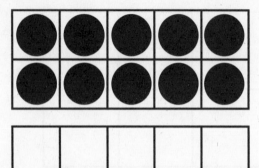

11

12

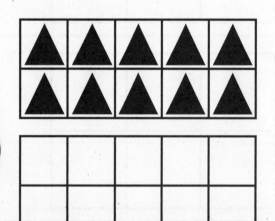

13

🐟 🐢 ★ Draw more shapes to show the number. Write the number.

© Harcourt

Name _____

14, 15, 16

🐟

14

15
🐢

16
★

🐟 🐢 ★ Draw more counters to show the number. Write the number.

© Harcourt

Practice PW 45

Name _____

Forward and Backward on a Number Line

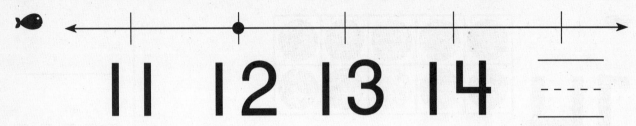

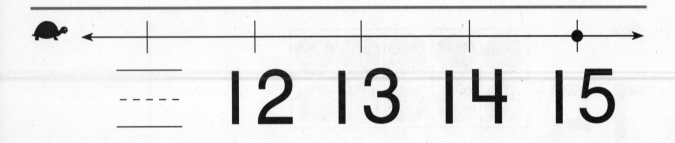

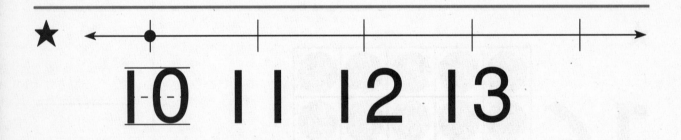

- 🐟 Start at 12. Count 3 numbers forward. Write the number.
- 🐢 Start at 15. Count 4 numbers backward. Write the number.
- ★ Start at 10. Count 4 numbers forward. Write the number.

17 and 18

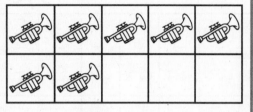 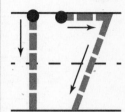

16   (17)   18

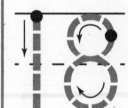

16   17   18

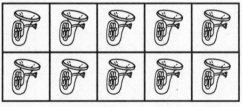 

16   17   18

_____
- - - - -
_____

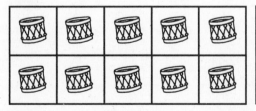 

16   17   18

_____
- - - - -
_____

16   17   18

🐟🐢 Count the instruments. Circle the number that tells how many. Trace the number.
★ ♥ Count the instruments. Circle the number that tells how many. Write the number.

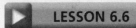
Problem Solving Skill: Use Data from a Graph

Toys

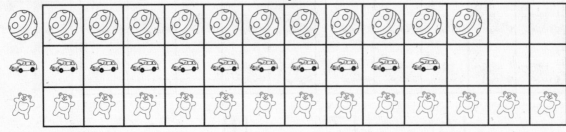

_____   _____   _____

 - - - - - - - - - - -   - - - - - - - - - - -   🐻 - - - - - - - - - - -

_____   _____   _____

Shapes

_____   _____   _____

♡ - - - - - - - - - -   ◇ - - - - - - - - - -   ☆ - - - - - - - - - -

_____   _____   _____

 Count and write how many. Circle the number that shows the most.
Mark an X on the number that shows the fewest.

Name _____

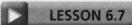

19 and 20

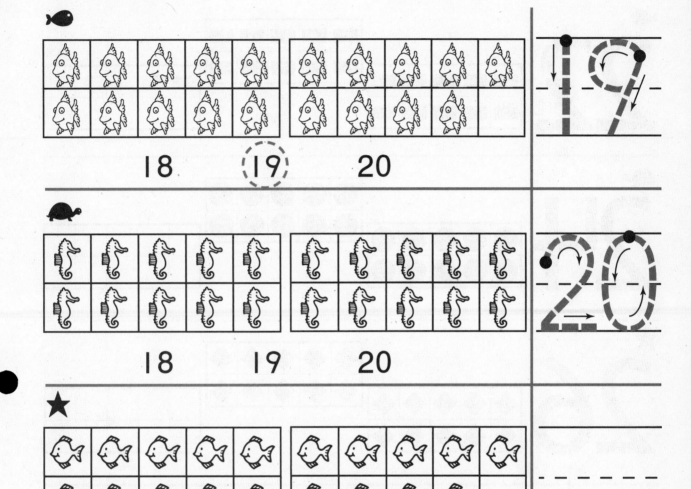

18   19   20

18   19   20

18   19   20

18   19   20

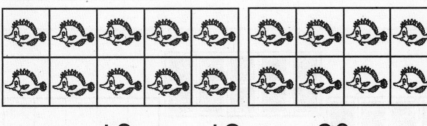
Count the fish. Circle the number that tells how many. Trace the number.
Count the fish. Circle the number that tells how many. Write the number.

© Harcourt

Name _____

21 to 30

🐟 22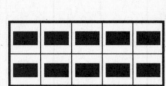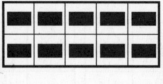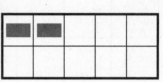

🐢 24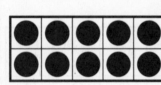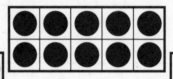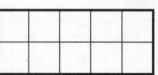

★ 26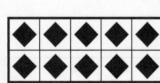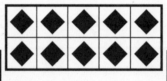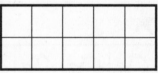

♥ 28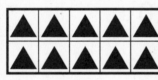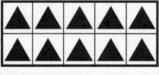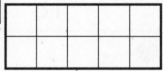

🌸 30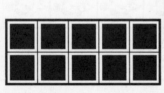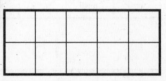

🐟 🐢 ★ ♥ 🌸 Draw more shapes to show the number.

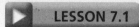

Name _____

Count Orally Using a Hundred Chart

(1)	2	3	4	5	6	7	8	9	10
11	12	13	14	15	16	17	18	19	20
21	22	23	24	25	26	27	28	29	30
31	32	33	34	35	36	37	38	39	40
41	42	43	44	45	46	47	48	49	50
51	52	53	54	55	56	57	58	59	60
61	62	63	64	65	66	67	68	69	70
71	72	73	74	75	76	77	78	79	80
81	82	83	84	85	86	87	88	89	90
91	92	93	94	95	96	97	98	99	100

Touch each number as you count from 1 to 40. Circle each number you say.
Touch and count from 41 to 80. Circle each number you say.
Touch and count from 81 to 100. Circle each number you say.

Name _____

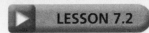

Count by 10s

1	2	3	4	5	6	7	8	9	10
11	12	13	14	15	16	17	18	19	20
21	22	23	24	25	26	27	28	29	30
31	32	33	34	35	36	37	38	39	40
41	42	43	44	45	46	47	48	49	50
51	52	53	54	55	56	57	58	59	60
61	62	63	64	65	66	67	68	69	70
71	72	73	74	75	76	77	78	79	80
81	82	83	84	85	86	87	88	89	90
91	92	93	94	95	96	97	98	99	100

Touch and count by tens. Use purple to color the numbers you say.

Count by 5s

5   10   15   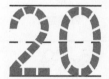

   30   35   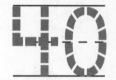

45   50   55   60

Place five connecting cubes above each flower. Count by fives. Trace the numbers.

Name _____

Count by 2s

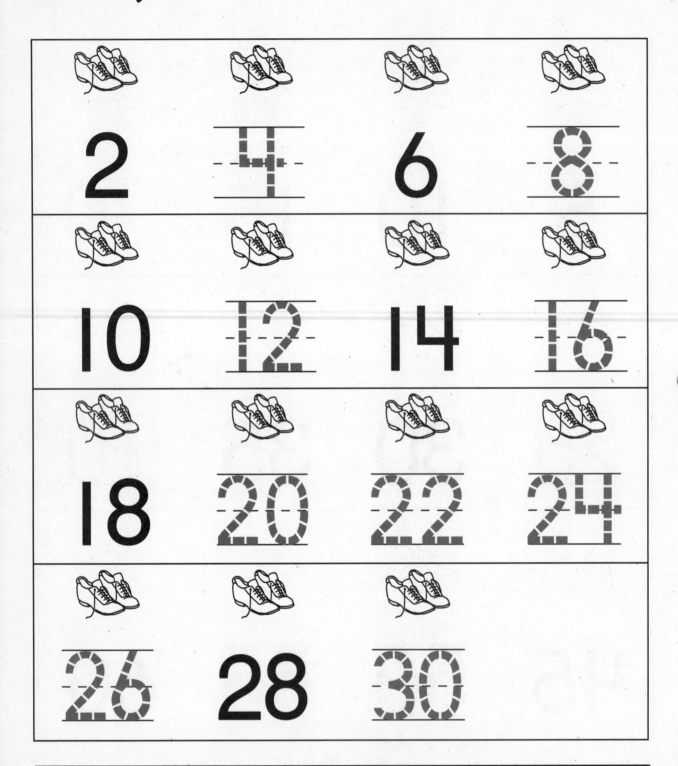

Count by twos. Trace the numbers.

Problem Solving Strategy: Find a Pattern

1	2	3	4	5	6	7	8	9	10
11	12	13	14	15	16	17	18	19	20
21	22	23	24	25	26	27	28	29	30
31	32	33	34	35	36	37	38	39	40
41	42	43	44	45	46	47	48	49	50
51	52	53	54	55	56	57	58	59	60
61	62	63	64	65	66	67	68	69	70
71	72	73	74	75	76	77	78	79	80
81	82	83	84	85	86	87	88	89	90
91	92	93	94	95	96	97	98	99	100

Use yellow to color the numbers you say when you count by twos. Use blue to circle the numbers you say when you count by fives. What pattern do you see?

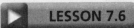

Counting by 10s

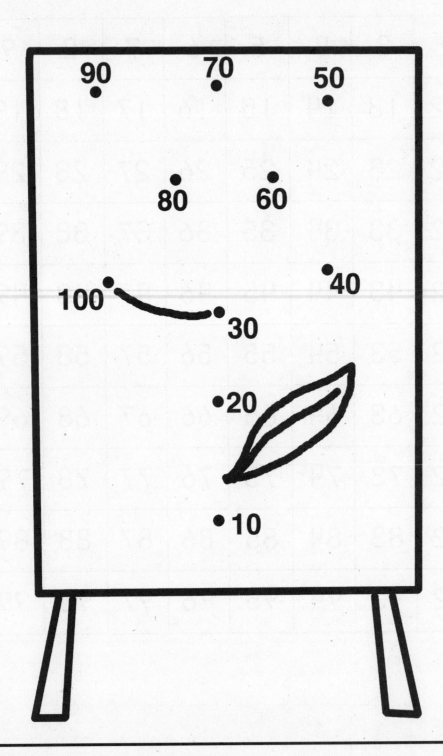

 Count by tens to connect the dots in order.

© Harcourt

Name _____

Even and Odd Numbers

(2)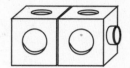

🐢

5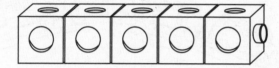

★

7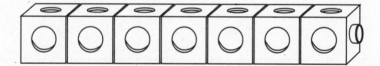

♥

9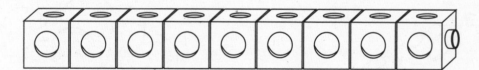

❀

10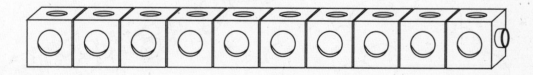

🐟 🐢 ★ ♥ ❀ Build the cube train. Take one cube from each end of the train and snap these cubes together. Do this as many times as you can. If you have only pairs, the number is even. Circle the number. If you have one cube left over, the number is odd. Mark an X on the number.

Name _____

Ordinal Numbers

first

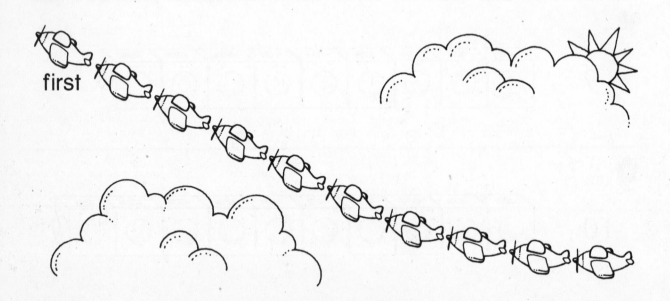

first

🐟 Circle the eighth bus. Draw a line under the third bus. Mark an X on the first bus.

🐢 Circle the fifth airplane. Draw a line under the tenth airplane. Mark an X on the fourth airplane.

Problem Solving Skill • Use a Model

🐟

4 5 6 7 8 9 10

🐢

4 5 6 7 8 9 10

★

4 5 6 7 8 9 10

🐟 Put your finger on 7. Count up 1. What number are you on? Circle the number.

🐢 Put your finger on 7. Count back 1. What number are you on? Circle the number.

★ Put your finger on 7. Count back 2. What number are you on? Circle the number.

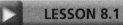

Name _____

Penny

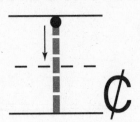 ¢

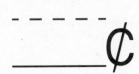 ¢

 ¢

¢

 Count the pennies. Write how many cents.

© Harcourt

Name _____

Nickel

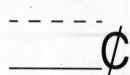

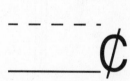

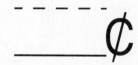

 Write how many cents.

© Harcourt

Name _____

— proper below —

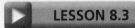

Dime

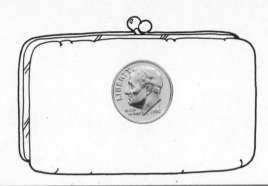

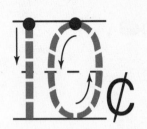

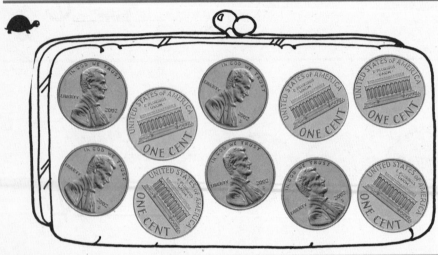

_____
- - - - -
_____ ¢

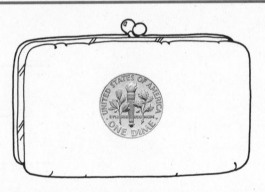

_____
- - - - -
_____ ¢

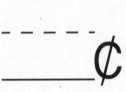

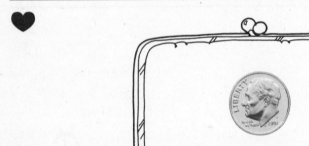

_____
- - - - -
_____ ¢

Write how many cents.

PW 62   Practice

Name _____

Morning, Afternoon, Evening

Morning

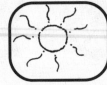
Afternoon

Evening

Morning

Afternoon

Evening

What times of the day do the pictures show? Circle the time of day that is missing.

Problem Solving Skill: Use a Calendar

October

Sunday	Monday	Tuesday	Wednesday	Thursday	Friday	Saturday
					1	2
3	4	5	6	7	8	9
10	11	12	13	14	15	16
17	18	19	20	21	22	23
24	25	26	27	28	29	30
31						

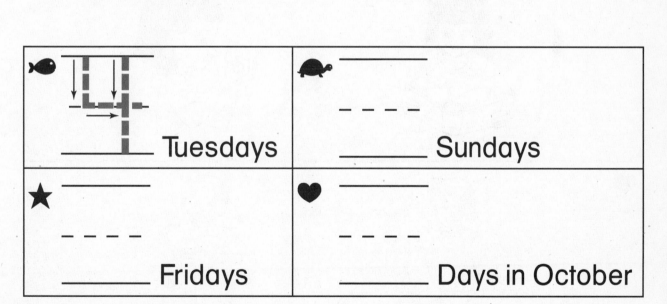

_____ Tuesdays   _____ Sundays

_____ Fridays   _____ Days in October

🐟 Count the Tuesdays. Write how many.
🐢 Count the Sundays. Write how many.
★ Count the Fridays. Write how many.
♥ Write how many days in October.

Name _____

More Time, Less Time

 Circle the activity that takes more time.

Name _____

Use a Clock

_____

_ _ _ _

_____ o'clock   _____ o'clock   _____ o'clock

_____

_ _ _ _

_____ o'clock   _____ o'clock   _____ o'clock

Write the number that tells the hour.
Circle the two clocks that show the same time.

© Harcourt

Name _____

Compare Lengths

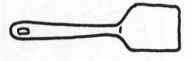

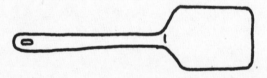

Circle the longer object. Draw a line under the shorter object.

PW68   Practice

© Harcourt

Order Lengths

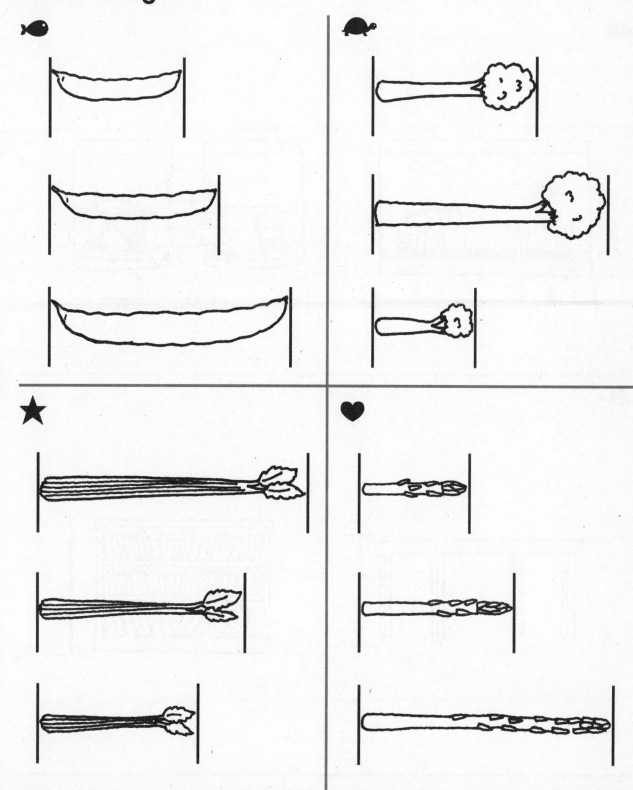

⊂⊃ 🐢 ★ ♥ Circle the groups of objects that are in order from shortest to longest, starting at the top.

Name _____

Indirect Comparison

   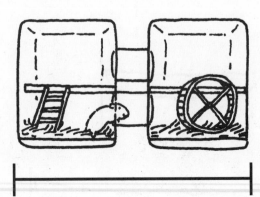

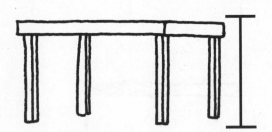   

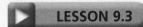 Go on a measurement walk. Use yarn of different colors to measure the two objects. Cut the yarn and compare the two pieces. Circle the picture of the place where you used the shorter piece.

© Harcourt

Measure Lengths with Nonstandard Units

_ _ _ _ _ _ _

_ _ _ _ _ _ _

★

_ _ _ _ _ _ _

♥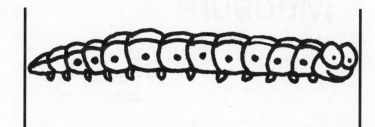

_ _ _ _ _ _ _

🐟 🐢 ★ ♥ Use cubes to measure the animal. About how many cubes long is it? Write the number.

Name _____

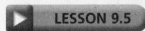

Problem Solving Strategy: Estimate and Measure

🐟 **Glue Stick**
50mL

Estimate

_____

- - - - - - - - - -

_____

Measure

_____

- - - - - - - - - - 🖇

_____

🐢

1   2   3   4   5

Estimate

_____

- - - - - - - - - -

_____

Measure

_____

- - - - - - - - - - 🖇

_____

★

Estimate

_____

- - - - - - - - - -

_____

Measure

_____

- - - - - - - - - - 🖇

_____

🐟🐢 ★ Estimate about how many paper clips long the object is. Then measure the object. Write how many paper clips long it is.

PW72 Practice

Compare Capacity

🐟 🐢 ★ ♥ Circle the container that holds more.
Mark an X on the container that holds less.

Name _____

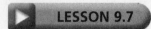

Compare Weight

 **Left**   **Right**

   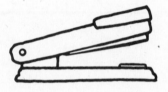

   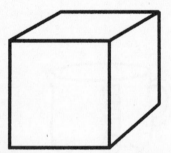

  ★   Hold one object in your left hand and one object in your right hand. Mark an X on the picture of the object that feels lighter.

© Harcourt

Problem Solving Skill: Use a Picture

★

 Use red to circle the pictures that most likely show hot weather.
Use blue to circle the pictures that most likely show cold weather.

Name _____

Make Concrete Graphs

🐟

How Many Blocks?

🐢

⬜

△

★ _____   _____

⬜ - - - - - - - - - -   △ - - - - - - - - - -

🐟 Place a handful of pattern blocks on the workspace.

🐢 Make a graph with your pattern blocks.

★ Write how many of each pattern block. Circle the number that shows the most blocks.
Mark an X on the number that shows the fewest blocks.

PW76 Practice

© Harcourt

Name _____

Read Picture Graphs

Which Snack Did You Eat?

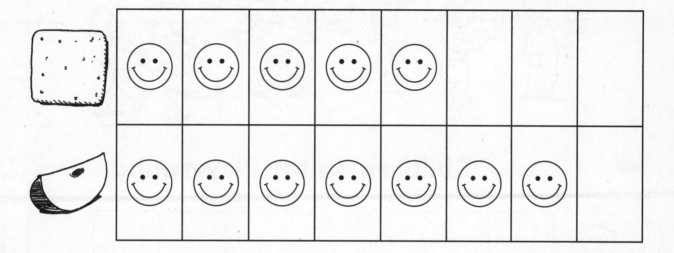

🐟 Use the graph. Write how many ate each snack.

🐢 Circle the snack that fewer children ate.

Name _____

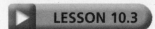

Make Picture Graphs

Children on the Playground

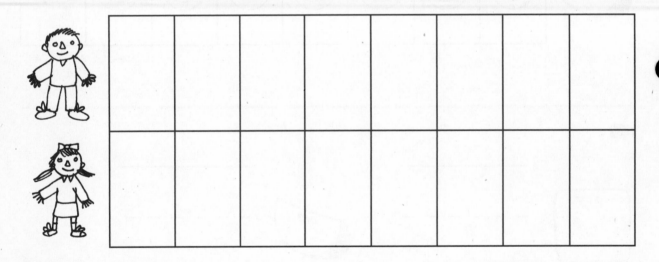

 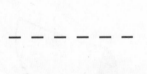  _____

-  Look at the picture. Make a picture graph about boys and girls.
- Write how many boys and how many girls. Circle the number that shows more.

PW78 Practice

Problem Solving Skill: Use Data from a Graph

Party Hats

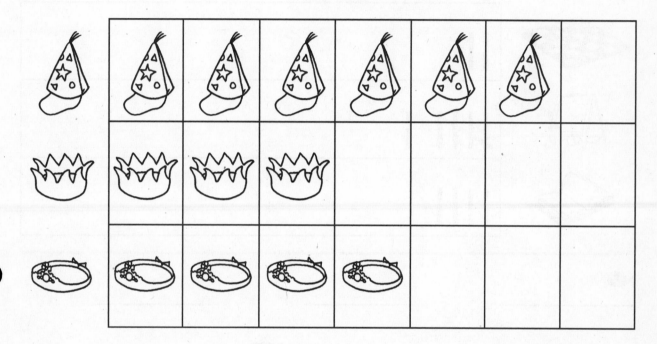

_____   _____   _____

- - - - -   - - - - -   - - - - -

_____   _____   _____

Look at the graph. Write how many of each kind of hat. Circle the number that shows the most hats. Mark an X on the number that shows the fewest hats.

Name _____

Read a Tally Table

Indoor Activities

	II
	IIII I
	IIII

 _____   _____   _____

★

 Look at the table. Write how many children do each activity.

🐢 Circle the picture that shows the activity the most children do.

★ Circle the picture that shows the activity the fewest children do.

Make a Tally Table

Do You Like Oranges?

🐟

Yes	
No	

🐢

Yes _____   No _____

🐟 Ask five classmates if they like oranges. Make a tally table.

🐢 Write how many for each answer. Circle the number that is greater. Do more of the children you asked like oranges or not like oranges? How do you know?

Name _____

Chance

Dark and Light

(dark)	
(light)	

Use a paper clip and a pencil to make a spinner. Spin ten times. Make a tally mark in the table after each spin. Which shade did the paper clip land on more often? Why did this happen? Circle the row with more tally marks.

PW82 **Practice**

Name _____

Explore Probability

Circle the picture that shows which is more likely to happen.

Circle the picture that shows which is less likely to happen.

© Harcourt

Problem Solving Skill: Make a Prediction

Ask ten classmates how many cartons of milk they drank at lunch.

Draw a picture to show how many cartons of milk in all.

Write how many cartons of milk the ten classmates drank.

Predict how many cartons of milk the whole class drank at lunch.

PW 84   **Practice**

Problem Solving Strategy: Act It Out

_____

- - - - - - - - - -

_____

- - - - - - - - - -

_____

   Tell an addition story about the picture. Act it out.
Write the number that tells how many children there are in all.

Model Addition

3   1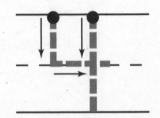

 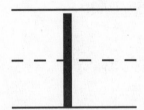

3   2   _____

2   2   _____

 Tell an addition story about the picture. Model the story with connecting cubes. Write the number that tells how many in all.

Name _____

Addition Patterns

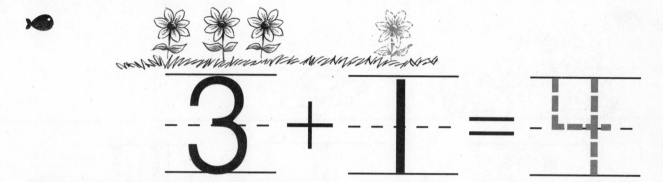

$$3 + 1 = 4$$

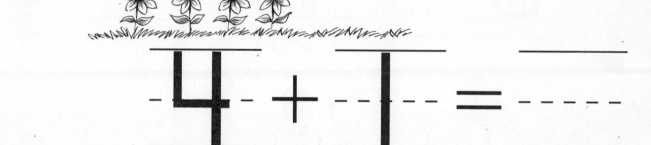

$$4 + 1 = \underline{\hspace{2cm}}$$

$$5 + 1 = \underline{\hspace{2cm}}$$

$$6 + 1 = \underline{\hspace{2cm}}$$

🐟 🐢 ★ ♥ Count the flowers. Then draw one more flower.
Write the number that tells how many flowers there are in all.

Name _____

Use Pictures to Add

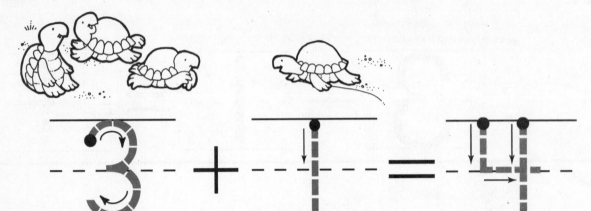

$$3 + 1 = 4$$

_____ + _____ = _____

★

_____ + _____ = _____

🐟 🐢 ★ Tell a story about the picture. Complete the addition sentence.

Name _____

Add With Money

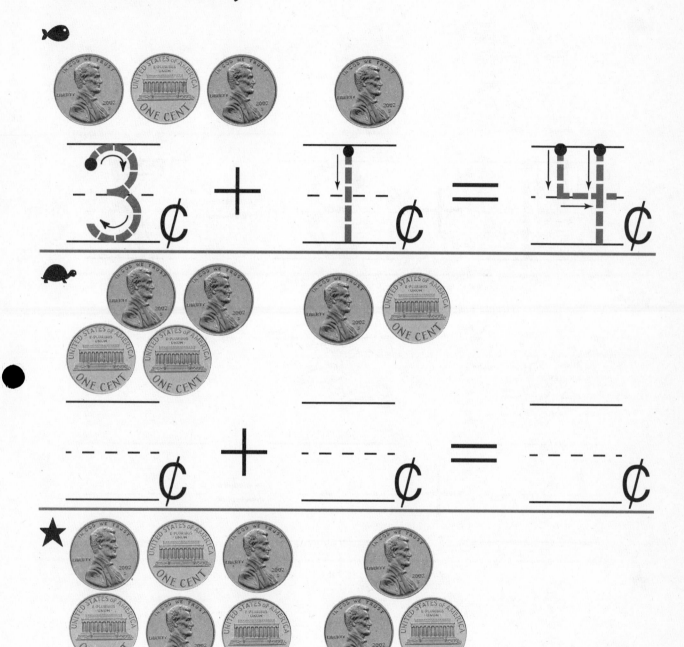

★ Count the pennies in each group. Write how many cents.
Add. Write how many cents in all.

Name _____

Addition Problems

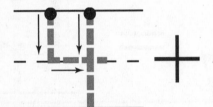 +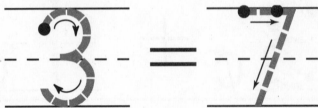

_____   _____   _____

- - - - - - + - - - - - = - - - - - - -

_____   _____   _____

★

_____   _____   _____

- - - - - - + - - - - - = - - - - - - -

_____   _____   _____

🐟 🐢 ★ Tell a story about the animals in the picture. Complete the addition sentence.

Addition Stories

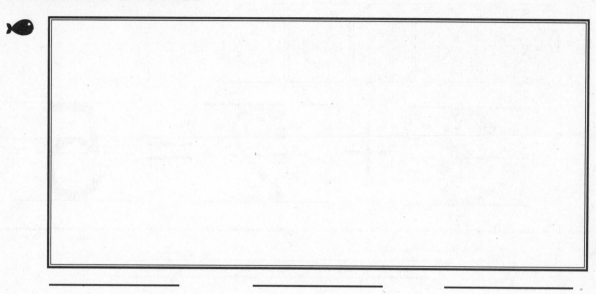

_____   _____   _____

$- - - - - \; + \; - - - - - \; = \; - - - - -$

_____   _____   _____

_____   _____   _____

$- - - - - \; + \; - - - - - \; = \; - - - - -$

_____   _____   _____

Tell an addition story. Model your story with objects.
Draw the objects and complete the addition sentence.

Problem Solving Strategy: Make a Model

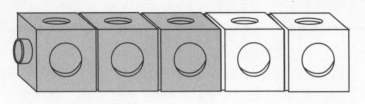

$$3 + 2 = 5$$

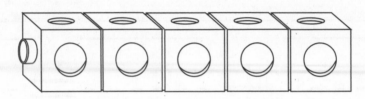

$$\underline{\quad} + \underline{\quad} = 5$$

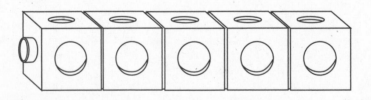

$$\underline{\quad} + \underline{\quad} = 5$$

🐟 🐢 ★ Use blue and yellow cubes to show different ways to make 5. Color the cubes. Write the numbers for each color to complete the addition sentence.

© Harcourt

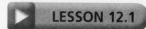

Problem Solving Strategy • Act It Out

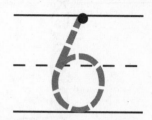

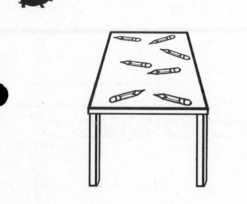

🐟 🐢 ★ Tell a subtraction story about the picture. Act it out. Count the objects that are left. Write the number that tells how many objects are left.

Name _____

Model Subtraction

6   4   _____
          - - - - - -
          _____

8   5   _____
          - - - - - -
          _____

Tell a subtraction story to go with the numbers. Model the story with connecting cubes. Write the number that tells how many are left.

Name _____

Subtraction Patterns

$$8 - 1 =$$ 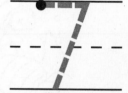 7

$$7 - 1 =$$

$$6 - 1 =$$

$$5 - 1 =$$

Count the seagulls. Mark an X on the seagull that is flying away.
Write the number that tells how many seagulls are left.

© Harcourt

Practice PW95

Name _____

Use Pictures to Subtract

$$8 - 4 = \underline{\hspace{2cm}}$$

$$9 - 3 = \underline{\hspace{2cm}}$$

🐟 🐢 Tell the subtraction story. Complete the subtraction sentence.

Name _____

Subtract with Money

$$5¢ - 2¢ = 3¢$$

_____ ¢ — _____ ¢ = _____ ¢

★

_____ ¢ — _____ ¢ = _____ ¢

🐟 🐢 ★ Tell a subtraction story. Complete the subtraction sentence to tell how much money is left.

Name _____

Subtraction Problems

 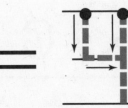

6 − 2 = 4

___ ___ ___
- - -   ___   - - -   =   - - -
___   ___

___ ___ ___
- - -   ___   - - -   =   - - -
___   ___

🐟 🐢 ★ Tell the subtraction story. Then complete the subtraction sentence.

PW98 Practice

Subtraction Stories

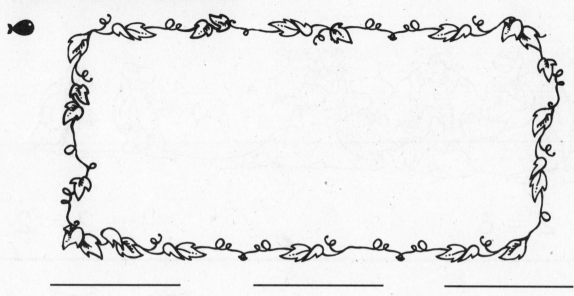

_____   _____   _____

- - - - - ▬▬▬ - - - - - ═══ - - - - -

_____   _____   _____

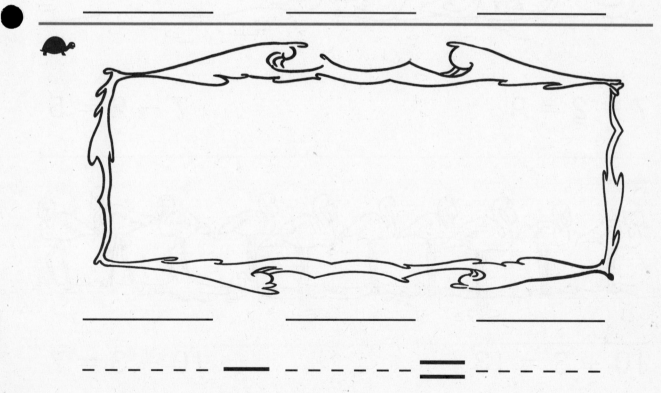

_____   _____   _____

- - - - - ▬▬▬ - - - - - ═══ - - - - -

_____   _____   _____

🐟 🐢 Tell a subtraction story. Act out your story with objects. Draw the objects, and mark an X on the objects you subtract. Complete the subtraction sentence.

Name _____

Problem Solving Skill • Choose the Operation

$4 + 2 = 6$ $4 - 2 = 2$

$7 + 2 = 9$ $7 - 2 = 5$

$10 + 3 = 13$ $10 - 3 = 7$

Tell a story about the picture. Circle the number sentence that shows what is happening in the picture.